MATHMAMIYA

MATHMAMIYA

LOVING MATHEMATICS AS A SECOND LANGUAGE

DEJI BADIRU

MATHMAMIYA
LOVING MATHEMATICS AS A SECOND LANGUAGE

iUniverse books may be ordered through booksellers or by contacting:

iUniverse
1663 Liberty Drive
Bloomington, IN 47403
www.iuniverse.com
844-349-9409

ISBN: 978-1-6632-6388-9 (sc)
ISBN: 978-1-6632-6389-6 (e)

Print information available on the last page.

iUniverse rev. date: 06/11/2024

Books in the ABICS Publications Book Series:
www.abicspublications.com

Also visit www.ABICS.com

1. **Mathmamiya: Loving Mathematics as a Second Language**, iUniverse, Bloomington, IN, 2024.
2. **Wreckless in the City: Physics of Safe Driving for young drivers (and adults too),** iUniverse, Bloomington, Indiana, 2024.
3. **Margin of Death: How close we come each day,** iUniverse, Bloomington, Indiana, 2024
4. **Soccer Greatness at Saint Finbarr's College (Volume II): Legacy of All-Around Sports**, iUniverse, Bloomington, Indiana, 2024
5. **Academics, Discipline, and Sports at Saint Finbarr's College: Tributes to the Great Soccer Players**, iUniverse, Bloomington, Indiana, 2023.

6. **More Physics of Soccer: Playing the Game Smart and Safe,** iUniverse, Bloomington, Indiana, 2022.
7. **Rapidity: Time Management on the Dot**, iUniverse, Bloomington, Indiana, 2022.
8. **The Physics of Skateboarding: Fun, Fellowship, and Following**, iUniverse, Bloomington, Indiana, 2021.
9. **My Everlasting Education at Saint Finbarr's College: Academics, Discipline, and Sports**, iUniverse, Bloomington, Indiana, 2020.
10. **Twenty-Fifth Hour: Secrets to Getting More Done Every Day**, iUniverse, Bloomington, Indiana, 2020.
11. **Kitchen Project Management: The Art and Science of an Organized Kitchen**, iUniverse, Bloomington, Indiana, 2020.
12. **Wives of the Same School: Tributes and Straight Talk**, iUniverse, Bloomington, Indiana, 2019.
13. **The Rooster and the Hen: The Story of Love at Last Look**, iUniverse, Bloomington, Indiana, 2018.
14. **Kitchen Physics: Dynamic Nigerian Recipes**, iUniverse, Bloomington, Indiana, 2018.
15. **The Story of Saint Finbarr's College: Father Slattery's Contributions to Education and Sports in Nigeria,** iUniverse, Bloomington, Indiana, 2018.
16. **Physics of Soccer II: Science and Strategies for a Better Game**, 2018.
17. **Kitchen Dynamics: The Rice Way**, iUniverse, Bloomington, Indiana, 2015.
18. **Consumer Economics: The Value of Dollars and Sense for Money Management**, iUniverse, Bloomington, Indiana, 2015.

19. **Youth Soccer Training Slides: A Math and Science Approach**, iUniverse, Bloomington, Indiana, 2014.
20. **My Little Blue Book of Project Management**, iUniverse, Bloomington, Indiana, 2014.
21. **8 by 3 Paradigm for Time Management**, iUniverse, Bloomington, Indiana, 2013.
22. **Badiru's Equation of Student Success: Intelligence, Common Sense, and Self-Discipline**, iUniverse, Bloomington, Indiana, 2013.
23. **Isi Cookbook: Collection of Easy Nigerian Recipes**, iUniverse, Bloomington, Indiana, 2013.
24. **Blessings of a Father: Education Contributions of Father Slattery at Saint Finbarr's College**, iUniverse, Bloomington, Indiana, 2013.
25. **Physics in the Nigerian Kitchen: The Science, the Art, and the Recipes**, iUniverse, Bloomington, Indiana, 2013.
26. **The Physics of Soccer: Using Math and Science to Improve Your Game**, iUniverse, Bloomington, Indiana, 2010.
27. **Getting Things Done Through Project Management**, iUniverse, Bloomington, Indiana, 2009.
28. **Blessings of a Father: A Tribute to the Life and Work of Reverend Father Denis J. Slattery**, Heriz Designs and Prints, Lagos, Nigeria, 2005.

Author Biography

Deji Badiru is a Professor Emeritus of Industrial and Systems Engineering. He is a registered Professional Engineer, a certified Project Management Professional, a Fellow of the Institute of Industrial & Systems Engineers, a Fellow of the Industrial Engineering and Operations Management Society, and a Fellow of the Nigerian Academy of Engineering. His academic background consists of BS in Industrial Engineering, MS in Mathematics, and MS in Industrial Engineering, and Ph.D. in Industrial Engineering. His areas of interest include mathematical modeling, systems engineering modeling, computer simulation, and productivity analysis. He is a prolific author and a member of several professional associations and scholastic honor societies. Deji holds a US Trademark for DEJI Systems Model for Design, Evaluation, Justification, and Integration.

Author Biography

[illegible]

Dedication

Dedicated to the memory of Professor Reginald Mazeres (February 15th, 1934 – December 15th, 2019), my extraordinary Professor of Mathematics at Tennessee Technological University, Cookeville, Tennessee. He used wily humor to make his class in Advanced Calculus fun, intuitive, and unforgettable. I still remember his lectures on Fermat's Last Theorem, Cantor Sets, and other advanced topics. Once again, I thank Ms. Michaela Rose Finn for her detailed copyediting work on this manuscript. With her at the helm of the copyediting chore, I could afford to be carefree in my initial composition of the manuscript, knowing that she would catch my glaring editorial blunders. Not much escapes her keen eyes.

Acknowledgements

I gratefully acknowledge all my past professors at Tennessee Technological University and University of Central Florida. Similarly, I appreciate the many years of collegiality and collaboration with my professorial colleagues at the University of Oklahoma, University of Tennessee, and the U.S. Air Force Institute of Technology. All are appreciated for the stimulating discussions and intellectual pursuits we engaged in. Mathmamiya is a celebration of the student mentoring I have engaged in for several decades.

WHAT'S IN A TITLE?

A rose by any other name is still a rose. A book by any other title is still the same book. The title of this book is based on a play on the Italian exclamatory words of "Mamma mia," which literally means "my mother, mom mine, mother of mine. This is idiomatically translated to mean "oh my mother, my goodness, oh my, etc."

"Mamma mia," can convey many different emotions including surprise, fear, pain, joy, and exasperation. Below is a nice original poem to help us solidify the premise of this book:

"Math, Math, Oh My Goodness.
Math, Math, What a Fear.
Math, Math, Oh My Mother.
How I wish you can unveil yourself to my understanding."
– Deji Badiru

The above sonnet suggests a "mamma mia" salutation.

But why did I use "miya" instead of "mia." That decision is from the influence of the Yoruba language of Nigeria, in which two vowels are not allowed in sequence. Vowels are interspaced by a consonant. This often creates a strong emphatic accent to the word in question. In this book, we have several influences of diverse spoken languages from around the world. This, indeed, makes mathematics a useable language that we need to love and adore. Hence, we have the book's title as *Mathmamiya: Loving Mathematics As a Second Language*. Mathematics isa language on its own. Now that you know the story behind this book's title, sit back and enjoy the fun of math at your fingertips. This book is written mostly from a comical and fun perspective to allay the fear of mathematics. Readers who are familiar with my previous writings may recognize the "Deji-Vu" pattern in my story-telling approach to teaching. It should be recognized that I tell inspirational and motivational stories because stories come with age. So, Deji-Vu stories continue. May the readers remain curious enough to read on and get motivated.

"When curiosity is established, the urge to learn develops." – Deji Badiru

My hope is that the presentations in this book will establish curiosity and interest to facilitate the urge to learn more about the fun of math. To this end, additional pertinent inspirational quotes can be found:

"Don't get frustrated, get motivated." – Deji Badiru

"Don't shy away from challenges; nothing develops a person better than adversity." – Deji Badiru

"Every new challenge is a learning opportunity."
– Deji Badiru

Mathematics, the abstract science of number, quantity, and space, has been tied to human existence since prehistoric times. The first recorded use of mathematics as a body of knowledge emerged in Babylonia in the third millennium, BC. It has since permeated every aspect of human civilization. It crosses subject boundaries from medicine to law, globally across nations. It is the key ingredient of any STEM (science, technology, engineering, and mathematics) career field. It affects all of us. So, we should all embrace and love it.

THE PURPOSE OF THIS BOOK

This is a math-mentoring book without the gory mathematical expressions. The intended purpose of this book is to allay the fear of mathematics that most people have. Math is actually a tame and fun ally in accomplishing a lot of things in life, such as investment, social networking, commerce, technology, engineering, and science. I encourage and spark the interest of young people, especially girls, to embrace Mathematics to their full potential. Consider the caption and quotes below:

Children lose confidence in their potential to "be scientists," but not in their capacity to "do science"

"Over the course of middle childhood, children's interest and beliefs about their own capacities for success in science often decline. *This pernicious decline is especially evident among underrepresented groups, including girls, members of some racial and ethnic minorities, and children from lower socioeconomic backgrounds.* The present research

(N = 306, ages 6–11) found that while children lose interest and feelings of efficacy about their potential to "be scientists" across middle childhood, they maintain more robust interest and efficacy about "doing science" …"

Source: Ryan F. Lei, Emily R. Green, Sarah Jane Leslie, and Marjorie Rhodes (2019), "Children lose confidence in their potential to "be scientists," but not in their capacity to "do science"." ***Developmental Science***, Vol. 22, e12837, Nov 2019 (**ERIC Number:** EJ1232329), *Wiley-Blackwell. 350 Main Street, Malden, MA 02148. https://ies.ed.gov/* *(*Institute of Education Sciences (IES) Home Page, a part of the U.S. Department of Education)

In a nutshell, the inherent ability may be there, but the confidence and pathway to manifest and leverage the ability may not be known. One way to discover and elicit the underlying ability is through humor and cajoling, which is the purpose of this freeform book.

This is <u>*not*</u> Math textbook, guidebook, or study guide. It is simply a fun book about the useful value of mathematics in our society and our daily lives, despite it being a "foreign language" to many of us. See the Calvin and Hobbes Comic that follows.

Many things around us in the world cannot be explained. However, mathematics can offer fun insights. Rather than being intimated by mathematics, we should learn to love it for what it offers us in terms of our basic human physical composition to the huge universe that governs our existence. Therein lies the theme of this book about "loving" and "having fun" with mathematics. This book is designed to inform, entertain, and encourage us about the role of mathematics in everything we do.

Please note that this book is not intended to teach mathematics, but rather to tease readers to whet the appetite for more math.

Happy reading!

Calvin and Hobbes Comic (Reprinted with permission)

WHY MATH MATTERS

Mathematics matters for a lot of things that we do, use, and enjoy. Front construction to space exploration, mathematics, either basic or complex, offers us insights into how things fit together. Rather than hating and avoiding it, we should embrace it for the greater good of the society.

The great Italian philosopher, Galileo Galilei (1564-1642), a polymath, long ago says "The book of nature is written in the language of mathematics."

Mathematics has various ramifications in the society and the advancement of humanity. Specific examples include the following:

- Number: This involves such things like counting, measurements of length, weights, and so on. It helps in displaying deep understanding of rational, real, complex, and p-adic numbers (in number theory).
- Shape: This leads to studies in geometries, topology, fractals, data mining, pattern recognition in artificial intelligence, and predictions.

- Movement: This involves activities of waves, planets, involving the use of ordinary and partial differential equations, Fourier Analysis, calculus of variations, and so on.
- Chance and Randomness: This involves the mathematics of probability, statistics, and stochasticity.

Shutterstock photo ID 1437782426. Reprinted with permission.

Math is all around us. It surrounds us, not only in the classroom, but also in the basic functions of life. Math is, indeed, a part of our human existence. The strength of numbers in everything we do ensures that we don't go overboard on many things. To some people, mathematics looks ugly and intimidating. To others, it is a thing of beauty to behold. I hope the comforting and humorous tone of Mathmamiya will help increase the math comfort level of readers.

Shutterstock photo ID 276131396. Reprinted with permission.

In 2016, when scientists announced the discovery of gravitational waves that have been traveling through the Universe for eons (specifically, for over 1.3 billion years), the calculations and projections were based on complex mathematics. I was so moved by the announcement that I wrote an educational article on "Beyond the Gravitational Jubilation: What the Discovery of Gravitational Waves Should Mean for Engineering Education." The article was published in the October 2026 issue of ***ASEE PRISM*** magazine of the American Society for Engineering Education. The gist of the article is replicated in the following narrative.

To the general public at that time, the exhilarating excitement and jubilation of the announcement beg the question of "so what?" in many circles outside the scientific communities. Even some researchers within the scientific groups were skeptical of the announcement, despite the worldwide multi-point scientific and mathematical confirmation. From an engineering educator's perspective, my focus was on what the eon-sourced waves could mean for the future of engineering education on Earth. Can we leverage the discovery to transform engineering education in ways that can help solve the nagging problems of the world? Even where there may not be a direct or apparent relationship to the theory of gravitational waves, can engineering education leverage the announcement along the spectra of motivating and inspiring new ways to energize engineering education? From my own engineering mind, the answers are, indeed, "yes" and "yes." We just have to look deeper beyond the immediacy of the discovery. After all, Albert Einstein theoretically predicated gravitational waves over one hundred years ago. Scientists have patiently dedicated time and space (pun intended) and Earthly resources to proving the existence of gravitational waves. Engineers and mathematicians should equally devote directed thinking at how the scientific breakthrough could influence how we craft and deliver engineering education of the future.

One thing that immediately comes to my mind is how the excitement should spark new appreciation and respect for the theory, mathematics, and natural laws of the Universe. What the Universe has fashioned eons ago is now reaching our own level of consciousness. So, the products of the Universe are pervasive and everlasting, even when we are not immediately conscious of the far-away developments. Our current world is just one tiny fleeting speck in the gamut of Universal occurrences. So, we should deliberate on how this realization can help us address our current worldly challenges. The still-unresolved 14 grand challenges for engineering (presented by NAE: National Academy of Engineering in 2008) come to mind as possible platforms for transforming engineering education based on the inspirational discovery of gravitational waves. Where there is no direct linkage, I believe there is a direct mathematical inspiration. The challenges and my own postulated connectivity with gravitational waves are listed below:

1. **Make solar energy economical**. Gravitational waves may be able to teach us new engineering ways to understand the travel of solar waves.
2. **Provide energy from fusion**. Gravitational waves may hold the key to the unknown energy potentials of fusion, from an engineering perspective.
3. **Develop carbon sequestration methods**. Gravitational waves may portend new atmospheric studies to advance carbon sequestration to mitigate, deter, deflect, or defer global warming.
4. **Manage the nitrogen cycle**. Gravitational waves may teach us new lessons of gas transportation with the nitrogen cycle.
5. **Provide access to clean water**. Can we study how water in the atmosphere is impacted by gravitations waves?
6. **Restore and improve urban infrastructure**. If we understand our place in the expanse of the universe,

we might be able to improve our long-term dedication to importance of urban infrastructure.

7. **Advance health informatics**. How the Universe preserves its information over the eons may teach us how we can design more sustainable health informatics.
8. **Engineer better medicines**. Is there a possible space-based platform for engineering better medicines?
9. **Reverse-engineer the brain**. This is a tough one, but our deep thinking about the universe may reveal how we doggedly pursue scientific questions.
10. **Prevent nuclear terror**. If we understand how gravitational waves permeate the Universe, we should have more fear of nuclear terror.
11. **Secure cyberspace**. Cyberspace cloud computing may one-day ride on the movements of gravitational waves.
12. **Enhance virtual reality**. When first postulated, gravitational waves were far-fetched. Now, they are a reality. This is a lesson of enhancing virtual reality.
13. **Advance personalized learning**. Learning about the universe of gravitational waves may influence how we dedicate ourselves to new personalized learning programs.
14. **Engineer the tools of scientific discovery**. Gravitational waves were discovered and confirmed by advanced scientific tools. Engineers should learn from this by engineering new tools for new scientific discoveries.

These are just my own conjectures as the author of Mathmamiya. Readers and educators out there are encouraged and invited to chip in with their own thoughts.

On a related note, mathematics facilitates exploration of the world around and beyond us. In 2021, I wrote another article on the collateral benefits of commercial space flight, which is being embraced more and more by the curious and the wealthy.

As former moon astronaut, Buzz Aldrin said, "We explore, or we expire." The emergence of commercial space flight, as evidenced by the space flight of Billionaire Richard Branson on July 11, 2021, portends a lot of collateral benefits for STEM (Science, Technology, Engineering, and Mathematics) education and practice. When equally thrill-seeking Billionaire Jeff Bezos followed suit with his own blast into suborbital space on July 20, 2021, my interest was further piqued by the potential for STEM opportunities created by these pseudo space adventurers. Public reaction was swift in creating two camps of sentiments. There are those who hailed commercial space flights as a welcomed technological advancement and there are those who saw the space show-off as a distraction from the social and economic ills plaquing our home-base Earth. While both the supporters and the detractors have their justifiable arguments, I see more of the pros than the cons. From a mathematical mindset with a systems viewpoint, I see more STEM opportunities to attract our future workforce. Detractors base their arguments on seeing the space pursuit by billionaires as a resource-wasteful and ego-driven competition at the expense of Earth-based human-focused needs. This is not a far-fetched position, considering the fact that we have several unresolved challenges here on Earth, including lack of adequate water, inability to eradicate disease, failure to eliminate poverty, social inequities, uncontrollable decline of the environment, as well as other pressing worldly challenges. My counter point is that the rapid execution of private commercial space flights will expedite the creation and dissemination of new engineering and technology products that may, very well, contribute to chipping away at our society's challenges.

The excitement of seeing a commercial vehicle blast into space and glide back to Earth is very inspiring for refocusing our STEM education topics into what can really help us advance the fundamentals of engineering beyond space flights. For example, many engineering students are terrified of undergoing lectures and tests in thermodynamics. As a motivation for learning more about the underpinning of what thermodynamics can do for us in practice on Earth, seeing the rocket blast-and-burn of Branson's SpaceShipTwo Unity (Virgin Galactic) can provide many motivational classroom presentations. Seeing is believing and that may be what more students require to re-dedicate their attention and efforts to the wide span of math-based education.

On the side of practical engineering practice, commercial space flights offer a multitude of multiplier effects affecting positive developments in business, industry, government, and the military. Collateral engineering businesses and commercial enterprises will, no doubt, develop to support the needs of routine commercial space flights. We should take the glass-half-full view of these positive developments rather shying away from new space pursuits.

Thinking back into history, what would have happened to the present world of air travel if the early curiosity of the Wright Brothers and others at the turn of the Century had been stymied by detractors who saw no value in flying from place to place around the world in a bird-like contraption? Similarly, in the early days of digital computers, some notable companies questioned why anyone would want a computer in his or her home. Now, we all benefit from the gregarious persistence of the early pioneers in air flight and home-based computing. Let's not miss an opportunity now to set a tone for the future of commercial space flight based on a love of the underpinning mathematics. We should encourage our present and future generations to love mathematics as a second language.

MATH AND MUSIC

It is often said that music and math go together. I believe it, not only because many musical symbols look like mathematical symbols, but because musicians are known to have analytical and mathematical minds. Although some musicians may possess latent math inclinations, they may never discover their math skills if they never get a chance to embrace and exercise their inner math potentials because of inherent fears. To be engaged (with Math) is to become comfortable (with Math). This book's approach of Mathmamiya may help remove the upfront block of fear (of Math) for many people. For example, as a state initiative in 2024, a case study in Tennessee, that included hundreds of Tennessee public schools, hypothesizes that music education may be tied to better performance in math and reading scores. Yes, music and math may be tied in ways that are not obvious until a math exposure occurs. My Mathmamiya may help in this regard.

MATH AND OTHER THINGS

Thomas Jefferson (1743-1826), the US third President said "Mathematics and natural philosophy are so peculiarly engaging and delightful as would induce everyone to wish an acquaintance with them. Besides this, the faculties of the mind, like the members of a body, are strengthened and improved by exercise. Mathematical reasoning and deductions are, therefore, a fine preparation for investigating the abstruse speculations of the law."

Along the same line, one Math advocate in the law profession at the turn of the 20th century was reported to have argued that "Our future lawyers, clergy, and statesmen are expected at the University to learn a good deal about curves, angles, number and proportions; not because these subjects have the smallest relation to the needs of their lives, but because in the very act of learning them they are likely to acquire that habit of steadfast and accurate thinking, which is indispensable in all the pursuits of life."

This implies that Mathematics is essential and applicable in all walks of life, even if we don't realize it right off the bat.

According to Underwood Dudley, a retired mathematician at DePauw University,

(Notices of the AMS, Vol 57 (5), 2010, 608-613) a contemporary student gave the following testimony: "The summer after my freshman year I decided to teach myself algebra. At school next year my grades improved from a 2.6 gpa to a 3.5 gpa. Tests were easier and I was much more e_ cient when taking them and this held true in all other facets of my life. To sum this up: algebra is not only mathematical principles, it is a philosophy or way of thinking, it trains your mind and makes otherwise complex and overwhelming tests seem much easier both in school and in life."

THE EARLY MATHEMATICIANS

My heroes are the early intellectuals of centuries ago, who created new bodies of knowledge that had never been seen before, documented, or even attempted. The intellectuals of that time did everything in science, mathematics, medicine, art, music, philosophy, and theology. They posed questions about the fundamentals of the nature of our universe and tackled those questions with dogged determination until they arrived at fundamental mathematical equations to explain what was, hitherto, unexplained. In many cases, the span of study stretched over many decades. They were not scared of what they didn't understand. They worked at each problem until a repeatable solution was discovered. One individual that always strikes me as an amazing intellectual was Sir Isaac Newton, an English polymath who was most famous for his laws of motion. He lived from 1642 to 1726 and contributed many other things to science and mathematics.

The most celebrated lessons about the physics of motion are conveyed by Newton's Laws of Motion, which he postulated in his 1687 book, *Mathematical Principles of Natural Philosophy.* Newton used the laws to investigate and explain the motion of many physical objects and systems. I have used the laws extensively on my Physics of Soccer website (www.physicsofsoccer.com), which highlights the transformative power of sports in education.

Newton was a mathematician, physicist, astronomer, alchemist, theologian, and author. He was described as a natural philosopher. He was an influential person in the scientific revolution and social enlightenment of his time. His pioneering book, *Mathematical Principles of Natural Philosophy* (1687), consolidated many previous results and established classical mechanics. Newton also made seminal contributions to optics, and shares credit with German mathematician Gottfried Wilhelm Leibniz for developing infinitesimal calculus, though he developed calculus years before Leibniz. Newton used his mathematical description of gravity to derive Kepler's laws of planetary motion, account for tides, the trajectories of comets, the precession of the equinoxes, and other phenomena. He demonstrated that the motion of objects on Earth and celestial bodies could be accounted for by the same principles. Newton's three famous laws of motion are summarized in the next section.

MATHEMATICS AND ARTIFICIAL INTELLIGENCE

You never know where and how Mathematics will lead us. In the 1970s, most math researchers thought the following two obscure pure math topics would never amount to anything. They were not expected to have any practical use beyond mere esoteric mathematical studies:

Prime number theory, which is now a critical part of cryptography in Internet operations.

Fourier-type transforms, which are now widely used and accepted in video compression processes.

With the way our digital world is advancing rapidly, we can never tell what branches of mathematics will provide the fundamental shifts in our new digital tools and techniques. I believe the future will be more and more mathematically driven, thus paving the way for more artificial intelligence developments and applications.

When I took Advanced Calculus in 1980 under Professor Reginald Mazeres at Tennessee Technological University, he introduced the topic of Cantor Set as an interesting mathematical study that did not have any practical applications. I fell in love with the Cantor Set's system of deleted middle third. Contrary to what Professor Mazeres said, I believe the set could, somehow, be forced into some practical engineering and science application. I spent several years, post-graduate studies researching and admiring the iterative process of deleting the middle thirds of the number line between 0 and 1. It was not until I started teaching in the College of Engineering at the University of Oklahoma in 1985 that I finally cracked the code of using Cantor Set to develop a search algorithm for artificial intelligence applications. My graduate students and I, subsequently, conducted research and published and/or presented several papers on the topic. Notable among them are the following:

1. Badiru, A. B., "A New Computational Search Technique for AI Based on Cantor Set," ***Applied Mathematics and Computation,*** Vol. 55, 1993, pp. 255-274.
2. Badiru, A. B., "Intelligent Computational Search Technique Using Cantor Set Sectioning and Modal Concentration," presentation at Partnership in

Technology Forum, BWXT Y12 National Security Complex, Oak Ridge, TN, February 28, 2003.

3. Milatovic, M. and A. B. Badiru, "Mode Estimating Procedure for Cantor Searching In Artificial Intelligence Systems," ***Proceedings of First International Conference on Engineering Design and Automation***, Bangkok, Thailand, March 1997.
4. Badiru, A. B. and L. Gruenwald, "A New Approach to Database Search Based on Cantor Set," ***Proceedings of International Conference on Computer Applications in Industry and Engineering***, Honolulu, Hawaii, December 15- 17, 1993, pp. 195-199.
5. Wei, Hong; L. Gruenwald; and A. B. Badiru, "Improving Cantor Set Search for Applications in Database and Artificial Intelligence," Proceedings of the 7th Oklahoma Symposium on Artificial Intelligence, Stillwater, Oklahoma, November 18-19, 1993, pp. 250-259.
6. Badiru, A. B., "Cantor Set Modeling for Manufacturing Knowledge Representation," presented at the ORSA/TIMS Spring Conference, New Orleans, April 1987.
7. Milatovic, Milan, "Development of Mode Estimating Technique for Cantor Search of Sorted Data in Artificial Intelligence Systems and Manufacturing Design," M.S. Thesis, School of Industrial Engineering, University of Oklahoma, 1996.
8. Nakada, Masayuki, "Experimental Investigation of Alternate Search Key Distributions for Cantor

Set Search Algorithm," M.S., School of Industrial Engineering, University of Oklahoma, 1993.

9. Wei, Hong, "Improving Cantor Set Search for Applications in Database and Artificial Intelligence," M.S. (Computer Science), University of Oklahoma, 1993 (co-chair with Dr. Le Gruenwald).

This is just another example of how mathematics can present new advancement opportunities in diverse STEM areas. Math is to be embraced, loved, and leveraged.

MATHEMATICS OF MOTION

I will use the sport of soccer to illustrate the mathematics of motion in this section. Mathematics govern motion, in accordance with Newton's Laws of Motion. Speed, velocity, and acceleration. These laws are especially important in sports. Speed is a scalar. Scalars are quantities with only magnitude. The direction does not matter. If someone is traveling on the interstate at 60 miles per hour (mph) south or 60 mph north, the speed is still 60 mph regardless. Other examples of scalar quantities are height, mass, area, and energy. Velocity is a vector. Both direction and quantity are important and must be stated. If one plane has a velocity of 400 mph north, and a second plane has a velocity of 400 mph south, the two planes have different velocities, even though the magnitude of their speed is the same. Both speed and velocity are important in playing soccer. Ordinarily, moving objects don't always travel with erratic and changing speeds, but in soccer that depends on the scenario of the game and directional changes of ball possession. Normally, an object will move at a steady rate with a constant speed. That is, the object will cover the same distance every regular interval of time. For instance, a marathon runner might be running with a constant speed in a straight line for several minutes. If the speed is constant, then the distance traveled every second is the same. If we could measure the runner's position (distance from an arbitrary starting point) each second, then we would note that the position would be changing by the same magnitude. This would be in contrast to an object which is changing its speed. An object with a changing speed would be moving a different distance each second. This is exactly what happens during a game of soccer. It is essential for young soccer players to have basic knowledge of science, technology, and mathematics to play soccer more intelligently by adapting to changes on the field.

Playing soccer is about motion, both of the player and the soccer ball. All motions are governed by the science of motion. In this respect, all soccer players must understand the basics of motion and how to use themeffectively for good game outcomes. Sir Isaac Newton, a 17th century scientist, developed laws to explain why objects move or don't move. His three laws of motion are popularly known as Newton's Laws of Motion: First Law, Second Law, and Third Law.

First law of Motion

Also known as the law of inertia, the first law of motion says:

"An object at rest tends to stay at rest and an object in motion tends to stay in motion with the same speed and in the same direction unless acted upon by an unbalanced force."

What is a force? A force is a directional push or pull upon an object resulting from the object's *interaction* with another object. Force is that which changes or tends to change the state of rest or uniform motion of an object. Whenever there is an *interaction* between two objects, there is a force upon each of them. When the *interaction* ceases, the two objects no longer experience the force. Forces exist as a result of an interaction between objects. There are two broad categories of force:

Contact Force

Action-at-a-distance force. Contact forces result from physical contact (interaction) between two objects. Examples include:

- Frictional force
- Tensional force
- Normal force
- Air resistance force

- Spring force
- Applied force

Action-at-a-distance forces result even when two interacting objects are not in physical contact with each other, but are still able to exert a push or pull on each other. Examples include:

- Gravitational force
- Electrical force
- Magnetic force

Force is a quantity which is measured using the standard metric unit known as the Newton (N). One Newton is the amount of force required to give a 1-kg (kilogram) mass an acceleration of 1 m/s/s (mass per second per second). Thus, the following unit equivalency can be stated:

1 Newton = 1 Kilogram mass/second/second.

This is written mathematically as $1\ N = 1kg(m)/s^2$.

A force is a vector quantity. A vector is a quantity which has both magnitude and direction. To fully describe the force acting upon an object, we must describe both the magnitude (size or numerical value) and the direction.

Second Law of Motion

Newton's second law of motion says:

"The acceleration of an object as produced by a net force is directly proportional to the magnitude of the net force, in the same direction as the net force, and inversely proportional to the mass of the object." This law is expressed by the following mathematical equation:

$$\text{Force} = \text{Mass} \times \text{Acceleration}$$

This written mathematically as $F = ma$.

F is the net force acting on the object, with a mass (m) and acceleration (a). This equation sets the net force equal to the product of the mass times the acceleration. Acceleration (a) is the rate of change of velocity (v) and velocity is the rate of change of distance.

Newton's first law of motion predicts the behavior of objects for which all existing forces are balanced. The first law states that if the forces acting upon an object are balanced, then the acceleration of that object will be 0 m/s/s. Objects at equilibrium (i.e., forces are balanced) will not accelerate. An object is said to be in equilibrium when the resultant force acting on it is zero (0). For three forces to be in equilibrium, the resultant of any two of the forces must be equal and opposite to the third force. According to Newton, an object will accelerate only if there is a net or unbalanced force acting upon it. The presence of an unbalanced force on an object will accelerate it, thus changing its speed, its direction, or both.

By comparison Newton's second law of motion pertains to the behavior of objects for which all existing forces are not balanced. The second law states that the acceleration of an object is dependent upon two variables: the net force acting upon the object and the mass of the object. The acceleration of an object depends directly upon the net force acting upon the object, and inversely upon the mass of the object. As the force acting upon an object is increased, the acceleration of the object is increased. As the mass of an object is increased, the acceleration of the object is decreased.

Third Law of Motion

Newton's third law of motion says:

> "For every action, there is an equal and opposite reaction."

Mathematically, a force is a push or a pull upon an object which results from its interaction with another object. Forces result from interactions between objects. According to Newton, whenever objects A and B interact with each other, they exert forces upon each other. When a soccer player sits in a chair, his body exerts a downward force on the chair, and the chair exerts an upward force on his body. There are two forces resulting from this interaction: a force on the chair and a force on the body. These two forces are called action and reaction forces in Newton's third law of motion. One key thing to remember is that inanimate objects such as walls can push and pull back on an object, such as a soccer ball. In ordinary parlance, it is a sort of tit for tat of the ball and the wall.

LEVERAGING HUMOR TO ALLAY FEAR

To most people, Mathematics is a foreign language. It often sounds like Greek to many. This is fitting because many of the symbols of mathematics have Greek origins. Many people will not touch Mathematics with a long pole, perhaps because they can't decipher the Greek. Mathematics offers many avenues of fun and intrigue. For example, the magic of numbers remains magical and unbelievable to the untrained eyes.

This is not a book about serious mathematics. Rather, it is a book of fun about the subject of mathematics. My belief is that approaching mathematics from a perspective of humor may help allay the fears of those who stay away from math, at least at the fundamental level. If we can use humor to pull back the curtain, mathematics may be revealed as a friendly companion.

My own love and intrigue with mathematics dates back to my high school days at Saint Finbarr's College, Akoka-Yaba,

Lagos, Nigeria. Based on my interest in the ordinary high school mathematics, I delved deeper into what was then called "additional mathematics." This was like the AP (advanced placement) mathematics courses in the USA high school system. At that time, the topics covered included Calculus. I found Calculus intimidating initially and contemplated dropping out of the course. However, a classmate, Babatunde Ogunde, who wasn't even taking the course, gave me a study book on "Add Math" (Additional Mathematics). The more I dug into the contents of the book, the more I loved Calculus. So, I stayed in the course, and even selected Additional Mathematics as one of my nine examinations for the 1972 West African Education Council (WAEC). I received a "Very Good" result in Mathematics and "Credit" in Additional Mathematics. This combination paved the way for my subsequent engagements with mathematics, not only in higher education, but also in professional pursuits.

From a humor angle, an engineer was heard saying "If there were no calculators, I would not study engineering. The arithmetic would be too cumbersome. With a calculator, engineering can be fun." There you have it. To this engineer, the basis of fun is the calculator. Which one is yours? I opine that it could be humor. To really enjoy mathematics, we must dispel the fear of mathematics upfront. Humor can put everyone at easy, with an enhancement of listening and paying attention to the fundamentals of mathematics.

In the community of my Nigerian high school days, the quadratic formula was given the nickname “Almighty Formula” to convey its efficacy in solving quadratic equations. That moniker scared off many students, believing that something was magical about the quadratic formula. Within a similar thread, I have many medical practitioner friends and family, who own express their detest of mathematics. These are extremely bright people, but they claim they would not get near mathematics with a ten-foot pole. This is based on an unfounded fear, which I hope to use humor to defuse in this book.

FINDING A MATHEMATICAL SOLUTION

It is inspirational to find a mathematical solution to a problem that was previously thought to be unsolvable. This brief section celebrates the mathematical accomplishment of two high school students, who recently added to the body of knowledge in developing a mathematical proof of the Pythagorean Theorem using Trigonometry.

In April 2024, newscaster Leila Sloman reported that two high school students proved the Pythagorean Theorem. At an American Mathematical Society meeting, these students presented a proof of the Pythagorean theorem that used trigonometry - an approach that some once considered impossible. The Pythagorean theorem allows you to calculate the longer side of a right triangle by summing the squares of the other two sides. These students proved the Pythagorean Theorem in a way that was once thought impossible by 20th-century mathematicians by using trigonometry.

Calcea Johnson and Ne'Kiya Jackson, both students at St. Mary's Academy in New Orleans, announced their achievement last month at an American Mathematical Society meeting. "It's an unparalleled feeling, honestly, because there's just nothing like it, being able to do something that people don't think that young people can do," Johnson told WWL-TV, a New Orleans CBS affiliate. If verified, Johnson and Jackson's proof would contradict mathematician and educator Elisha Loomis, who stated in his 1927 book *The Pythagorean Proposition* that no trigonometric proof of the Pythagorean theorem could be correct. Their work joins a handful of other trigonometric proofs that were added to the mathematical archives over the years. Each sidestepped "circular logic" to prove the pivotal theorem. So, what exactly is a trigonometric proof of the Pythagorean theorem, and why was Loomis so closed off to the idea?

(https://www.scientificamerican.com/article/2-high-school-students-prove-pythagorean-theorem-heres-what-that-means/.

The Pythagorean Theorem provides an equation to calculate the longer side of a right triangle by summing the squares of the other two sides. It is often phrased as $a^2 + b^2 = c^2$. In this equation, a, b and c represent the lengths of the three sides of a right triangle, a triangle with a 90-degree angle between two of its sides. The quantity c is the length of the longest side, called the hypotenuse. Though the theorem is named for the ancient Greek philosopher Pythagoras, some historians believe it was known in Babylon around 1,000 years earlier.

The theorem "connects algebra and geometry," says Stuart Anderson, a professor emeritus of mathematics at Texas A&M University–Commerce. "The statement $a^2 + b^2 = c^2$, that's an algebraic statement. But the figure that it comes from is a geometric one."

Meanwhile trigonometry focuses on functions that depend on angles. These functions, such as the sine and cosine, are defined using right triangles. Imagine a right triangle with one side that lies flat against a table and another that shoots straight up from where it meets the first side at a right angle. The hypotenuse will reach diagonally between these two sides.

Now measure the angle between the hypotenuse and the table. Mathematicians define the sine of this angle as the height of the vertical side divided by the length of the hypotenuse. The cosine of this angle is the length of the horizontal side divided by the hypotenuse. The Pythagorean Theorem is therefore equivalent to the equation $\sin^2 x + \cos^2 x = 1$. "A lot of the basic trig 'identities' are nothing more than Pythagoras' theorem," explains Anderson, referring to equations that describe relationships among different trigonometric functions.

MATH AND STATISTICS

Math and Statistics often go hand in hand. There are no statistics without math. The information echoed below, and the job prospects are culled from my alma mater's Math newsletter, *The Radical Times* (Vol. 17, No. 1, September 2023, page 2), under the heading of "COOL MATH JOBS."

Currently, statisticians are one of the most popular mathematical professions. Statisticians are considered crucial for the success and advancement of any business or industry. Statisticians play a vital role in research, policy development, risk assessment, and business optimization. Although the specific tasks that statisticians complete vary depending on the industry and organization they work for, all statisticians have the same objective: to use statistical methods, mathematical theories, and models to solve real-world problems by collecting, analyzing, interpreting, and presenting data.

Since the objective is the same for all statisticians, some responsibilities are universal. Most statisticians have the following duties: to provide evidence-based solutions and informed strategies by identifying patterns, trends, and relationships within data and drawing meaningful conclusions from data. These conclusions are used to design experiments, surveys, and data collection methods as well as use statistical models that describe and predict behavior or outcomes based on the data. They also present their findings and analysis of results in a clear and understandable manner through charts, graphs, and reports, perform data validation, outlier detection, and data cleaning to eliminate errors and inconsistencies in the data. Finally, they complete or learn about research that develops new statistical methods, improves existing techniques, and contributes to the advancement of the field of statistics.

Similarly, all statisticians are expected to possess analytical skills, problem-solving skills, logic and reasoning skills, technical skills, communication skills, and leadership skills. Statisticians often use computer programming languages and software to perform various aspects of their job such as analyzing data sets, performing complex calculations, and developing statistical models. Likewise, it is common for statisticians to collaborate with colleagues and professionals from different disciplines, as well as present their findings to stockholders and/or colleagues. Therefore, statisticians need to communicate in a clear and accessible way. Furthermore, leadership skills are important since statisticians need to understand the objectives of their clients or collaborators.

Some common types of statisticians include Academic Statistician, Biostatistician, Econometrician, Environmental Statistician, Financial Statistician, Government Statistician, Machine Learning Statistician, Market Research Statistician, Quality Control Statistician, Social Statistician, and Sports Statistician. Statisticians are employed in most industries. Typically, statisticians that work in the private sector will collect, analyze, and interpret data in order to provide informed organizational and business strategies, whereas in the public sector they often focus their efforts on furthering the public good.

In 2021, Statisticians had a mean salary of $95,570 and the job outlook is expected to grow 31% by 2031. Education requirements for becoming a statistician can vary depending on the company or industry. Usually, there is a requirement of a bachelor's degree in mathematics or a related field. However, some require a master's degree in mathematics or even a PhD.

MULTI-DIMENSIONAL MATH CALLS

In this section, a call is an invitation to participate and learn math through a variety of educational programs. There are a slew of approaches online, in books, in videos, and in classrooms devoted to enhancing the teaching of Math. Math is more reachable than detractors realize. The key is to find the best way to teach Math to the various targeted groups effectively. Some of the approaches entail cajoling, encouraging, and monetizing (as in scholarships). There are **Mathmania** programs designed to encourage kids to embrace and learn Math. There are **Mathmarathon** games to improve Math fluency in a fun setting. There is the **MathCount** competition (http://mathcounts.org/) with a focus on competitively enhancing Math knowledge. The city of Dayton, Ohio annually hosts the **Dayton TechFest** science fair, where Math is a primary focus. In 2010-2012, I organized a booth at TechFest to promote using Math and Science to improve sports performance. My website for that educational outreach, www.PhysicsOfSoccer.com, is still active, up, and running. The central feature of the website is my 2010 book entitled, ***The Physics of Soccer: Using Math and Science to Improve your Game***. My mathmamiya presentation in this book adds to the multi-dimensional approaches in promoting the embrace of mathematics.

The Ohio Montgomery County Educational Service Center (MCESC) regularly organizes workshops and Math strategy programs for teachers. One of their summer programs is organized to help teachers work on multisensory strategies for teaching math concepts. In June 2023, teachers from the Dayton region met for a two-day training about using multisensory math strategies to teach math concepts to students. The multisensory approach to teaching math uses visual, structural, and conceptual ways of learning, as opposed to using memorization and repetition. I recall that in my own elementary school days, we were drilled on memorization and repetition. That approach probably disengaged many kids, who could have, otherwise, been math whizzes. Math is a science of patterns and relationships, as we see throughout the theme of this book. The flexibility offered by the more contemporary technique multisensory strategies allows teachers to effectively convey new math knowledge to students. The more kids (and adults too) that can be nudged to visualize numbers, the more they can develop math problem-solving strategies. Mathematics has tremendously large and diverse applications. It should be understood, enjoyed, and utilized by more people in the population. This is the premise and promise of this book.

MATH ATTENTION AND LEARNING

I am convinced that listening is a fundamental ingredient for appreciating and learning Math. In this section, I recount a testimony from my own teaching of mathematics as a graduate teaching assistant at Tennessee Technological University, Cookeville, Tennessee in 1979. If you have heard or read about this testimony before, it is okay, we will just add this as an incident of my famous "Deji-Vu Stories." The more stories are re-told, the more they are likely to stick and the more readily the message can be conveyed and retained.

On the preceding note about listening and learning, I once had a real-life experience with a student's epiphany with Algebra. It was 1979 and I was a freshly minted industrial engineering graduate from the College of Engineering at Tennessee Technological University. Intent on furthering my curiosity about mathematics, I enrolled in graduate school in the Department of Mathematics at the same school. With my engineering education and good grades in my undergraduate math courses, it was easy for me to secure a graduate teaching assistant job in the department. This was the first time I ended up in front of a classroom, teaching undergraduate Algebra, having just graduated from being on the opposite side of a classroom. As a new and young instructor, I leveraged my inherent humor to get the attention of the students. We all had fun while a retainable transfer of knowledge transpired. All went well and all were equitably rewarded with their respective end-of-quarter grades in accordance with their course performance. We were on an academic quarter system in those days, rather than the present-day semester system. Students were required to complete end-of-quarter instructor evaluations, in which, thankfully, I did well, with many students citing my relaxed classroom atmosphere, driven by my inspirational humor. One particular written testimony that captured my attention was a statement by a student, who, obviously, had gotten the exciting flavor of Algebra. She wrote:

"I came into this class trembling, knowing that I might flunk algebra again. I have flunked the course four previous times and I still could not get it. This is my last opportunity to take the course, which is required for my graduation. If not absolutely required, I would have stepped away from Algebra. I just couldn't get it. On the first day of class, this instructor, Deji Badiru, came into the classroom and I was definitely sure I would flunk again. When regular American instructors taught the course, I flunked. I believed that this new instructor with a funny accent that I could hardly understand, my flunking was inevitable. But being my last chance, I decided to give it my best shot. I decided to always sit in the front row so as to better hear and understand this instructor. Not only that, I paid a close attention to everything he said. With my extra attentiveness, I was amazed how well I understood his lectures and the fun of Algebra struck me. I fully expect to pass the course this time. Thank you, Deji Badiru."

Of course, by university policy, I did not see this testimony until after the course ended and grades were assigned and recorded officially. I was delighted that this student, who shunned animosity and proudly appended her name and signature, did end up with a grade of A for the course. So, to me, this confirms that humor could be a good conveyance of knowledge.

MATH AND THE ADVANCEMENT OF EDUCATION

Some of the greatest fun of mathematics is right around our corner, in our neighborhood, and in our everyday lives. A great joy can be derived when Math is applied in the community, which affects everyone of us. In his 2019 TEDxDayton discussion (https://www.youtube.com/watch?v=E5ID4F3gwGs), Dr. Luther Palmer says this best with his assertion that his love affair with math was "love at first fraction." Dr. Palmer, who claims to be a math addict, emphasizes that one of the beauties of math is its objectivity. He recollects that his mother used to say, "One plus one is always two regardless of your skin color or what part of town you're from." This is true because math is consistent and predictable if all the input parameters are known andcorrect. Dr. Palmer goes on to note that there is more to math than its objectivity. He explains why the context in which math is done matters as much as the answer it provides. This statement is very true. Unfortunately, the more a person enjoys math, the less likely he or she will think about the context and problems to which their math skills will be applied. Dr. Palmer concludes that "Math is objective in its doing, but it's anything but objective in its using." In this context, the avowed minister of Math, who is a Professor of Electrical Engineering at Wright State University in Dayton, Ohio, has dedicated himself to spreading the good news about the transformative and life-saving power of Mathematics and Science. He enjoins his listeners to step back from Math to reflect and appreciate its beauty and promise. His Math crusade matches the "Mathmamiya" theme of this book as well as the advocacy presented in my Physics of Soccer website (www.PhysicsOfSoccer.com), the homepage of which is echoed in the image below:

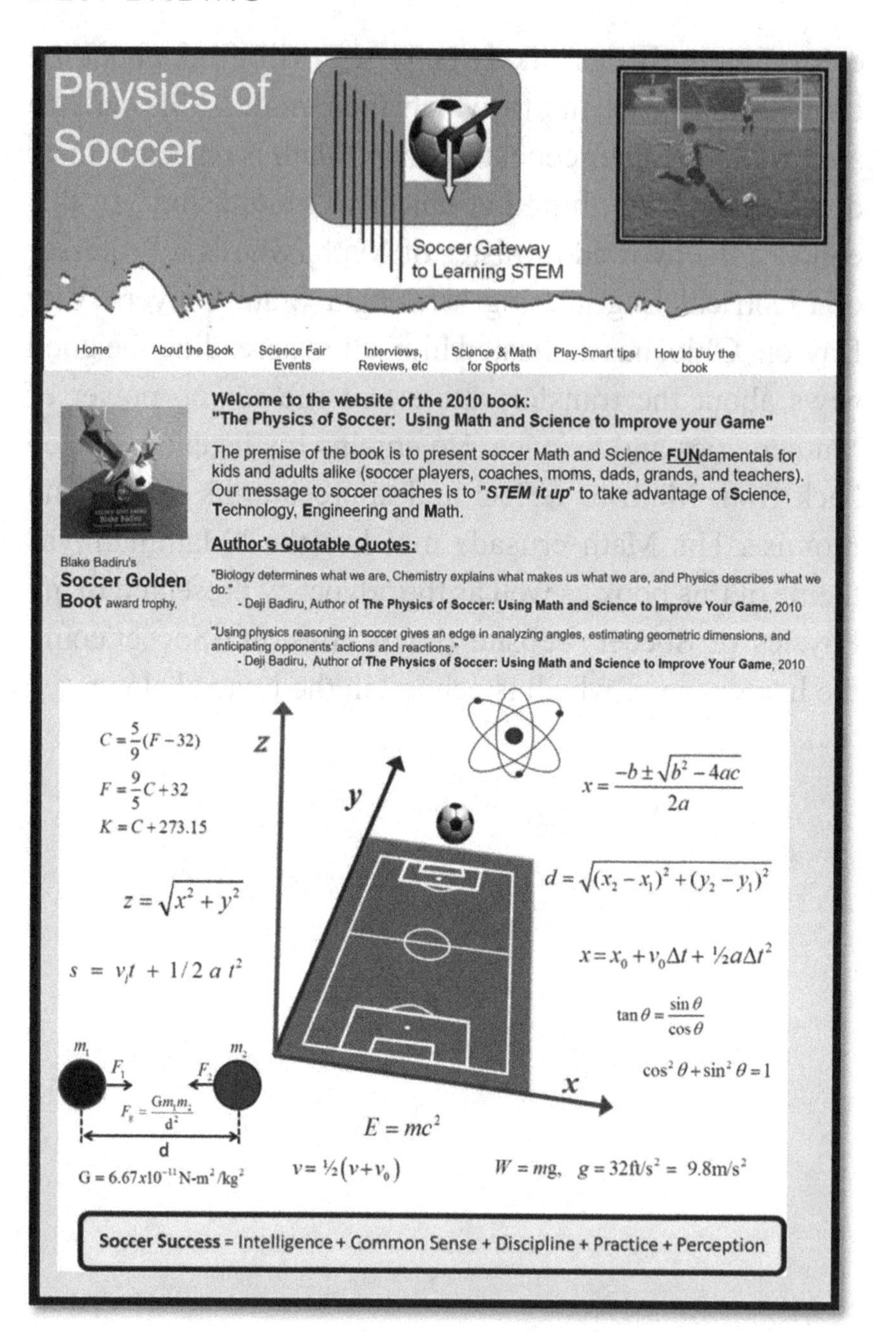

(Author's physics of soccer website image)

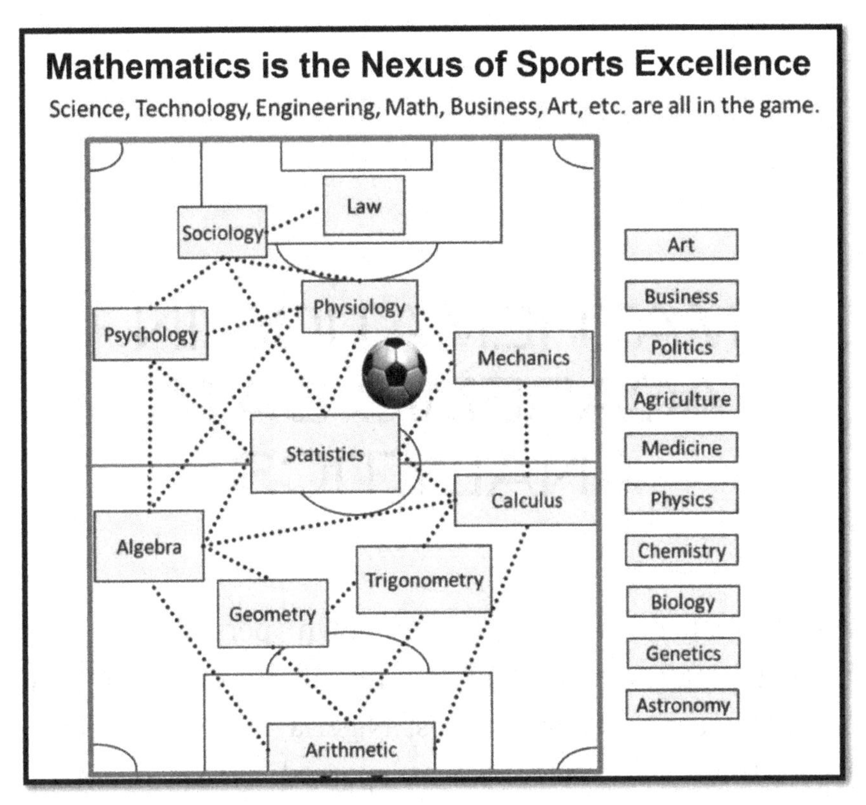

(Author's physics of soccer website image)

Numerous other applications of mathematics in the community can be found both online and in the published literature.

MATHEMATICS IN ACADEMIC DATA ANALYTICS

As previously mentioned, my own personal interest in mathematics dates back to my high school days at Saint Finbarr's College, Lagos, Nigeria in the late 1960s. Interestingly, that interest was spurred by the historical accounts of how Florence Nightingale used mathematics and statistics as a tool in promoting a crusade for hygiene throughout the population. One of our teachers at Saint Finbarr's College was particularly enamored with Florence's historical exploits and he lectured repeatedly about her accomplishments to the extent that I developed an interest in how math and statistics could be used for personal planning. Florence Nightingale's data-driven practices were credited with significantly improving hospital efficiency of that era. I was very intrigued. Florence Nightingale was a British social reformer and statistician, as well as the founder of modern nursing. Nightingale came to prominence while serving as a manager and trainer of nurses during the Crimean War

(October 1853–February 1856), in which she organized care for wounded soldiers. With an interest in boosting my academic performance in selected subjects, I started developing hand-drawn mathematical charts to do a trend analysis of my course grades. It was an interesting application that caught the attention of my classmates and teachers. With the trend charts, I could see the fluctuations in my course grades, based on which I intensified my study habits either to maintain or improve outcomes in specific courses. It was a rudimentary application of visual data analytics for a personal topic of interest, although the name "data analytics" was far from what I would have called it, it worked. I did not think much of that basal tool of course-grade statistics until I started my industrial engineering studies at Tennessee Technological University many years later. As of today, I still have the original hand-drawn data analytics trend chart of my high school course grades from 1968 through 1972. Although faded in resolution, it still conveys my long-standing interest and dogged pursuit of using mathematics and data analytics. I can say there were many "Mathmamiya: moments conveyed by my rudimentary charts, as can be seen in the chart below.

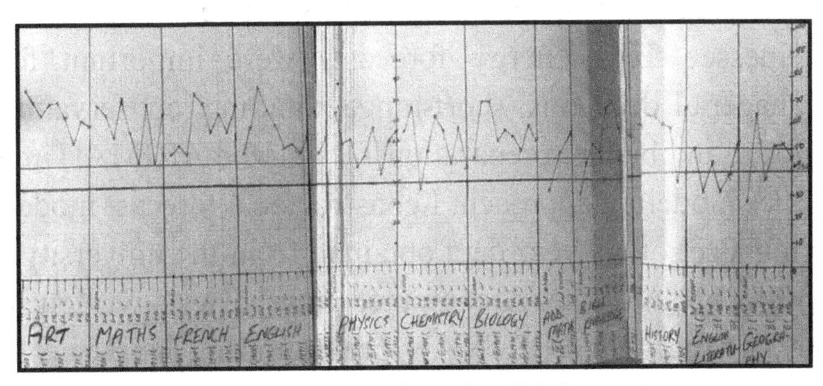

(Author's original data plot JPEG image)

The figure presents an archival illustration of what my 1972 chart looks like. My mathematics and additional course grades can be seen on the chart. It is like taking a step back in data-visualizing time. It is important to note that the figure is a scanned reprint of the original hand-drawn chart from 1968 to 1972. The importance of preserving the original appearance of the chart, albeit in low-resolution print quality, is to convey that data visualization does not have to be esoteric, as long as the mathematical background is preserved. Even a rudimentary hastily hand-drawn chart may provide sufficient information for corrective or proactive actions to enhance a desired end result. The chart in the figure, as poor in print quality as it may appear here, provided the desired impact over the grade-tracking years of 1968 through 1972. In 1981, I completed a master's thesis research on mathematical modeling of energy consumption at Tennessee Technological University. The thesis research concerned the development of a stochastic forecasting model for energy consumption at the university. Such a model is useful for obtaining reliable forecasts of future energy consumption. This was an important pursuit in that era because of the energy crisis between 1979-1980. Tennessee Tech energy forecasts were important for managerial decisions, short-range planning, conservation efforts, and budget preparations. Using the ARIMA Time Series modeling approach, I constructed a forecast model, using a set of six-year data obtained from the university's physical plan office. A computer program was written in the now-defunct FORTRAN programming language to achieve the model building and the forecasting processes. Once again, mathematics played a role in personal and organizational performance. Several mathematical models and variations of data analytics are used in practice. Below are some of the more common models:

1. Conceptual models
2. Theoretical models
3. Observational models
4. Experimental models
5. Empirical models
6. Iconic models
7. A-priori reasoning

The conceptual data model is a structured business view of the data required to support business processes, record business events, and track related performance measures. This model focuses on identifying the data used in the business but not its processing flow or physical characteristics.

A theoretical model can be defined as a theory that is developed to explain a situation or a phenomenon and be able to forecast it. Theoretical modeling is based on a number or a set of theories. These theories are used to explain some situations, phenomena, behavior types.

This term is more commonly used in qualitative research, while the term theoretical model usually appears as a tool in quantitative research. They both refer to the key theories, models, and ideas that exist in relation to your chosen topic. They give your research a direction and set boundaries for the reader. Observational learning is a type of learning that happens indirectly through a process of watching others and then imitating, or modeling, their behavior. A 1961 Bobo doll experiment demonstrated how school-aged children modeled aggressive behavior seen in adults. Experimental model is an example of conducting experiments to collect data to develop understanding that can then be transferred to another situation of interest. A common example is the use of animal experiments to model the human case. Animals that are employed to model the development and progression of diseases, and to test new treatments before they are given to humans, are used in experimental modelling. Animals with transplanted human cancers or other tissues are called xenograft models. Empirical modelling refers to any kind of computer modelling based on empirical observations rather than on mathematically describable relationships of the system modelled. Empirical modelling is a generic term for activities that create models by observation and experiment. Empirical Modelling (with the initial letters capitalized, and often abbreviated to EM) refers to a specific variety of empirical modelling in which models are constructed following particular principles. An iconic model is an exact physical representation and may be larger or smaller than what it represents. The characteristics of an iconic model and the object that it represents are the same. This area is now popularly known as "digital twin" methodology in Industry 4.0 applications. It is frequently used in Operations Research analysis. In a priori reasoning, the modeling is based on knowledge or inclination originating from theoretical deduction rather than from observation, experience, or empirical data assessment.

MATHEMATICS OF HAPPINESS

As I stated earlier, Math has wide and diverse areas of application. As a fun application, I present my mathematical formulation of happiness, which we all desire. Nowadays, we live in a computationally-intensive society, predicated on digital manufacturing. The Crypto-Currency industry readily comes to mind. It helps to understand some of the basic mathematical principles and tools that underpin everything we do.

Mathematical Fun with Factors of Happiness

Thinking mathematically, we can define happiness as a function of several factors (or variables). Behind the factors, there are indicators and attributes of happiness. In this regard, we can quickly get complex by defining a multi-tiered mathematical equation of happiness. Fortunately, we won't get that ambitious in this fun formulation. So, we will keep the equation very basic as presented below:

$$H = \mathrm{f}(x, y, z, \text{etc.})$$

H represents happiness

f represents the mathematical function. Hopefully, a simple algebraic expression. I know some of our readers will be looking for something more excitingly complicated. Thankfully, they are in the minority. We will tone things down for the benefit of the majority.

x, y, z represents the factors to be included in the definition of happiness.

etc. represents any additional string or series of factors of interest.

Below is a list of some possible factors affecting happiness:

- Your Artwork
- Your Bed
- Your Boat
- Your Books
- Your Car
- Your Church
- Your Cooking
- Your Drinks
- Your Faith
- Your Family
- Your Fellowship
- Your Food
- Your Freedom
- Your Friends
- Your Game
- Your Health
- Your House
- Your Job
- Your Kitchen
- Your Leisure
- Your Lineage
- Your Love
- Your Money
- Your Marriage
- Your Neighborhood
- Your Pets
- Your Reading
- Your Religion

- Your School
- Your Shoes
- Your Space
- Your Sports
- Your Stocks
- Your Wardrobe
- Your Worship
- Your Writing

Fortunately, life is not so complicated that we need to include too many factors of happiness. For most people, it takes just a handful of factors to define their acceptable level of happiness. For our illustration here, let's use only five factors. We know, at the extreme, there may be those who define their happiness solely on one factor. Different strokes for different folks, so to speak. The factors we will be using include: Health, Wealth, Family, Fellowship, and Writing. Readers are encouraged to think of and rationalize about their own five factors. Now, we can go back and define our equation of happiness as follows:

$$H = f(Health, Wealth, Family, Fellowship, Writing)$$

Now, not all factors are created equal. To account for this nonuniformity, we assign numbered relative weights to the factors. So, our equation becomes:

$$H = f(w_1 Health, w_2 Wealth, w_3 Family, w_4 Fellowship, w_5 Writing)$$

The w_i is the relative weight of each respective factor. The factors, each ranging between 0 and 1.0, can be equal or variegated depending on their individual levels of importance or desirability. The requirement is that the weights all sum up to 1.0. That is, in the usual mathematical convention, the weights should comply with the following:

$$w_1 + w_2 + w_3 + w_4 + w_5 = 1.0$$

This ensures that we weight each factor in relation to the other factors with respect to the level of importance. Note that different people may rate these five factors differently, but with each being between 0 and 1.0, keeping the sum as 1.0. To make mathematicians happy, we will write this as follows:

$$\sum_{i=1}^{N} w_i = 1.0$$

N represents the number of factors included in the happiness function. In our example, N = 5.

Now, we need to apply the functional form of our equation. Again, we will keep it simple, as a basic additive function. Thus, we end up with the following, as a simple-minded example:

$$H = w_1(\mathit{Health}) + w_2(\mathit{Wealth}) + w_3(\mathit{Family}) + w_4(\mathit{Fellowship}) + w_5(\mathit{Writing})$$

If we now ask each person to rate each factor on a normalized independent scale of ***Desirability*** (D) between 0 to 100, we can come up with a numerical score for each factor, against which each relative ***Weight*** (*w*) will be applied, so that we can calculate an overall measure of happiness (H), based on the selected five factors. For a computational example, let's assume the following distribution of weights and desirability of the factors. Note that each weight cannot be bigger than one and each factor cannot be bigger than 100.

Factor	**Weight**	**Desirability**	**(w)(D)**
Health	0.35	100	35.00
Wealth	0.20	95	19.00
Family	0.25	100	25.00
Fellowship	0.15	85	12.75
Writing	0.05	65	3.25
TOTAL	**1.00**	445	**95.00**

Thus, the composite numeric happiness level for this hypothetical person is 95. This would be compared and evaluated against other combinations of factors, weights, and desirability ratings. Technically, happiness computations should not be compared across individuals, since the component elements are expected to be different for different people. This computational exercise may just be used as a fun way of determining the level of satisfaction with what constitutes happiness for each person. In this example, we could very well set the Desirability of each factor at the highest level of 100, since they are independently rated. That would indicate that all the factors are equally desirable at the highest possible level. I expect the students among us to toy around and do the toil of playing with the above formulation, if only for the purpose of just having fun.

MATHEMATICS OF RULE OF 72

Mathematics plays a crucial role in economics and finance, which affects everyone. Analysis of investments, incomes, and gains can be tracked mathematically. The Rule of 72 is a simple mathematical relationship that is often used in personal finance

The Rule of 72 is an easy way to estimate how long it will take for an investment to double, given a fixed annual interest rate. By dividing 72 by the annual rate of return, you can get a rough estimate of the number of years it will take to double your initial investment.

This rule is a quick way to understand the impact of compound interest. Compound interest is the principle by which the interest you earn also earns interest. It's an important investment concept, and it's why you should continually reinvest your earnings. Notice how much more you can earn by holding investments that compound as opposed to those that don't.

For comparative purposes, we could apply the Rule of 72 to two different investments to determine which one will double in the shortest amount of time. The one that doubles first would earn the most in the long run.

The concept of interest has existed since ancient Mesopotamian, Greek, and Roman civilizations. It's also mentioned in the Quran. It was primarily used in agriculture and land transactions, as well as standard money loans. Lenders have always needed to know how much they'll earn from an investment.

The first reference we have of the Rule of 72 comes from Luca Pacioli, a renowned Italian mathematician. He mentions the rule in his 1494 book *Summa de arithmetica, geometria, proportioni et proportionalita* (*Summary of Arithmetic, Geometry, Proportions, and Proportionality*). Here's what Pacioli says about it:

"In wanting to know of any capital, at a given yearly percentage, in how many years it will double adding the interest to the capital, keep as a rule [the number] 72 in mind, which you will always divide by the interest, and what results, in that many years it will be doubled. Example: When the interest is 6 percent per year, I say that one divides 72 by 6; 12 results, and in 12 years the capital will be doubled."

THE MAGIC OF NUMBERS

Numbers have fascinating and almost magical properties that we should leverage for encouraging the embrace of mathematics. If it is fun, students (and others) will be interested in learning more.

In case you missed it earlier, I am purposely repeating my earlier case example of teaching algebra to spark the interest of students in general mathematics. One case example is my use of fun analogies and jokes about numbers to heighten the excitement and interest of my students when I taught College Algebra many years ago as a Graduate Teaching Assistant in the Department of Mathematics at Tennessee Technological University in Cookeville, Tennessee in 1979. A student's course evaluation comment indicated that she passed the College Algebra course on the basis of her attentive interest and listening to my funny classroom lecture style. She got a Grade of A. According to her end-of-course evaluation, she had failed the same course four times prior, until she came to take my class. She reported that she was very enthused by my practical explanations and real-world analogies. She listened with rapt attention to everything that I had to say in the class. The end result was that she understood the basic Algebra concepts that she had not previously grasped. Therein lies the power of Mathmamiya, as conveyed in this book. Get the student's attention and the student will get the concept. My expectations are to get the reader's attention here so that they can easily embrace mathematics, which is a desirable outcome for this educational outreach. This reminds me of the quote below:

> "Teachers don't teach for the income.
> Teachers teach for the outcome."

Algebra is often feared by college students, but it happens to be a fun subject with many practical applications around us. We just have to find a way to get students interested and attentive. That's the primary purpose of my Mathmamiya presentations in this book.

One funny joke about the fear of mathematics is conveyed in a student's letter to Algebra. The letter says:

==========================

"Dear Algebra,

Please stop asking us to find your X.
She is gone. We don't know where she is. We believe she is not coming back.

We don't know where Y is either. We think they are together somewhere.

Thank you for understanding our plight of not knowing where X and Y are."

==========================

Clever letter, indeed.

All kinds of number patterns exist to tickle our fancy. Consider the pattern of numbers and prefixes below.

yotta (10^{24}):	1 000 000 000 000 000 000 000 000
zetta (10^{21}):	1 000 000 000 000 000 000 000
exa (10^{18}):	1 000 000 000 000 000 000
peta (10^{15}):	1 000 000 000 000 000
tera (10^{12}):	1 000 000 000 000
giga (10^{9}):	1 000 000 000
mega (10^{6}):	1 000 000
kilo (10^{3}):	1 000
hecto (10^{2}):	100
deca (10^{1}):	10
deci (10^{-1}):	0.1
centi (10^{-2}):	0.01
milli (10^{-3}):	0.001
micro (10^{-6}):	0.000 001
nano (10^{-9}):	0.000 000 001
pico (10^{-12}):	0.000 000 000 001
femto (10^{-15}):	0.000 000 000 000 001
atto (10^{-18}):	0.000 000 000 000 000 001
zepto (10^{-21}):	0.000 000 000 000 000 000 001
yacto (10^{-24}):	0.000 000 000 000 000 000 000 001
Stringo (10^{-35}):	0.000 000 000 000 000 000 000 000 000 000 000 01

If numbers appear enthralling, interest and curiosity will follow and further learning can develop.

Magic Star

A magic star is an n-pointed star polygon, in which numbers are placed at each of the n vertices and n intersections, such that the four numbers on each line sum to the same magic constant. A normal magic star contains the integers from 1 to 2n with no repetitions. This type of mathematical construct has interesting properties that can glue the attention of students and adults. The magic star is a variation of magic squares and other geometric patterns.

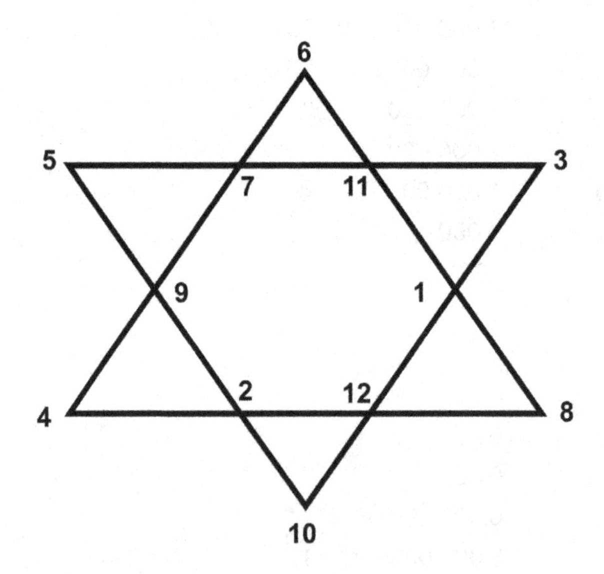

Magic Hexagram

As stated previously, a normal magic star contains the integers from 1 to 2n, with no numbers repeated. The magic constant of an n-pointed normal magic star is $M = 4n + 2$. In the example below, $n = 6$ (pointed edges). $M = 4(6) + 2 = 26$. Readers are encouraged and invited to verify the sum of numbers along each straight line. No star polygons with fewer than 5 points exist. Thus, the construction of a normal 5-pointed magic star is said to be impossible. Consequently, the smallest example of a normal magic star is 6-pointed. Never say never. As shown by the 2023 accomplishment of Calcea Johnson and Ne'Kiya Jackson in finding a new trigonometric proof for Pythagora's Theorem, which had been thought to be impossible for centuries, and may turn out to be achievable through determination and persistence. A reader here may very well be the first to find a new magic star with less than 6 points.

Readers are encouraged to entertain themselves by exploring other interesting patterns offered by other similar geometric patterns including magic circle, magic hexagon, magic hexagram, magic square, magic star, magic triangle, alphamagic square, antimagic square, geometric square, heterosquare, prime reciprocal magic square, most-perfect magic square, magic cube, magic hyper cube, magic hyperbeam, associative magic square, pandiagonal magic square, multimagic square, word square, number scrabble, eight queens puzzle, multimagic square, magic constant, and magic series. The math fun continues in all dimensions.

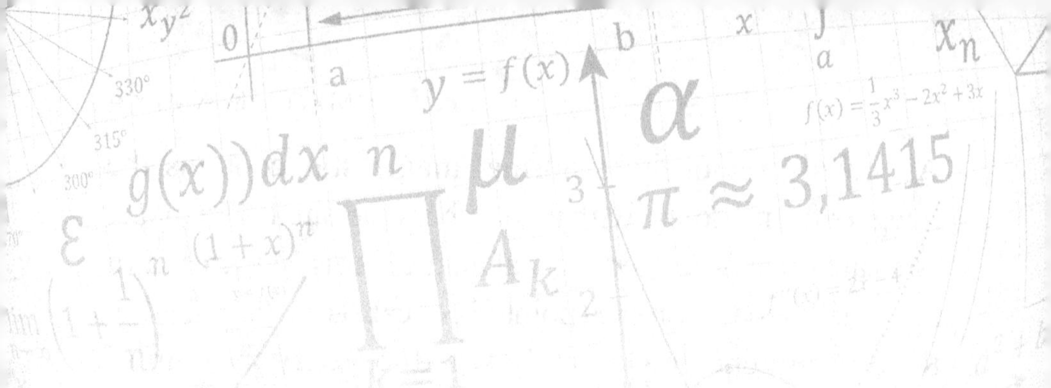

MAGIC OF SQUARING

There is fun and magic in squaring any number in the range of 90 to 99. Below is how it is done.

Mathematical expression of the Magic:
$(100 - a)^2 = 100 \times [(100 - a) - a] + a^2$

Steps:

1. Find the difference between the number and 100.
2. Square the result of step 1 and write it down (make sure it takes up a two-place digit). For example, 3 is written as 03 and 7 is written as 07.
3. Subtract the result of step 1 from the original number and write it down in front of the number in Step 2. That is the result of the square.

Example 1 (Find the square of 97)
$97^2 = ????$
1) 100 – 97 = 3.
2) $3^2 = 9$. Write 09 to take up two place values.
3) 97 – 3 = 94. Write 94.
The answer is 9409.

Example 2 (Find the square of 92)

$92^2 = ????$

1) $100 - 92 = 8$.
2) $8^2 = 64$. Write 64.
3) $92 - 8 = 84$. Write 84.

The answer is 8464.

Very interesting and entertaining. There are scores, if not hundreds of such fun games of numbers.

IMAGINARY NUMBERS

Imaginary numbers are numbers that, when squared, have a negative result. They are also the square root of negative numbers. Imaginary numbers are written as real numbers multiplied by the imaginary unit "i", where $i = \sqrt{(-1)}$. Imaginary numbers are not real numbers, meaning they cannot be represented on the number line, but they are used in math to solve complex equations. They can also be comically used in complex tax computations, as jokingly represented in the LOCKHORN comic below.

LOCKHORNS Comic Strip: Reprinted with permission

MAGIC NUMBER NINE

Multiply nine by any whole number other than zero, then repeatedly add the digits of the answer until you get a single digit. That digit will always be the number nine. Very interesting, magical, and inviting to try.

Example 1: 9 x 1,357 = 12,213
1 + 2 + 2 + 1 + 3 = 9

Example 2: 9 x 8,642 = 77,778
7 + 7 + 7 + 7 + 8 = 36

Now, from 36, we have 3 + 6 = 9.

Readers are invited to play around with this and try new examples.

MAGIC NUMBER EIGHT

1. Write down your phone number, followed by the number 8.
2. Multiply your phone number by 8 and write the answer down next to the 8 after your phone number.
3. Add all the single digits together until you only have a single digit left. What is it?

Example: 12345… → 1+2+3+4+5+… = ????

GENERAL MACARTHUR'S NUMBER GAME

1. Write down the number of the month you were born.
2. Double it.
3. Add 5.
4. Multiply by 50.
5. Add in your age.
6. Subtract 365.
7. What is your number?

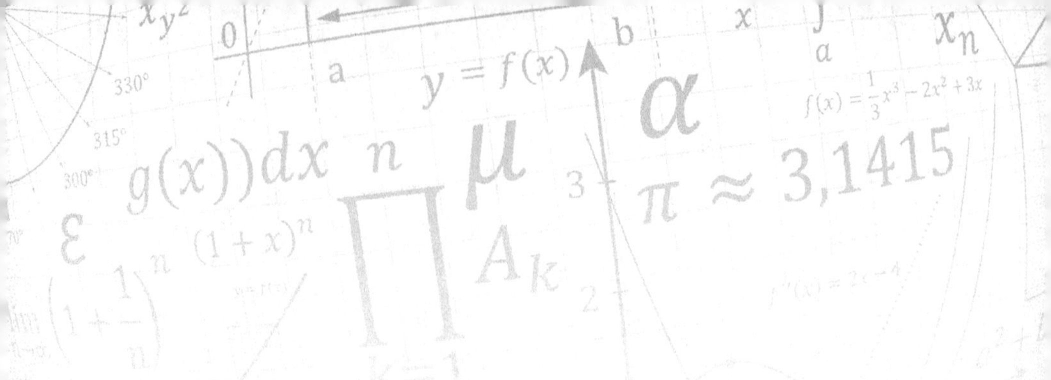

THE INTRIGUE OF 2520

One of the basics of mathematics is knowing whether a number is even or odd, divisible by 2 and by the sum of its numbers. The number 2520 is very strange in this regard. It looks like an ordinary and a normal number. But it is not. It is a strange number that has puzzled mathematicians for centuries. Only the number 2520 is divisible by numbers from 1 to 10, regardless of whether they are odd or even. The division is clean. That means there are no fractions remaining from the division. Not many numbers have that property. Below are the division examples of 2520.

$2520 \div 1 = 2520$

$2520 \div 2 = 1260$

$2520 \div 3 = 840$

$2520 \div 4 = 630$

$2520 \div 5 = 504$

$2520 \div 6 = 420$

$2520 \div 7 = 360$

$2520 \div 8 = 315$

$2520 \div 9 = 280$

$2520 \div 10 = 252$

The strangest thing yet is that this number 2520 is a product of 7 x 30 x 12. This is not a random coincidence. 2520 = 7 days per week x 30 days in a month x 12 months in a year. Very intriguing, indeed. Of course, naysayers will ask about leap years, 31-day months, and February. Well, have math fun your way.

SOCCER MATH JUST FOR KICKS

Mathematics plays direct roles in sports. This is the genesis of my 2010 book entitled *The Physics of Soccer: Using Math and Science to Improve Your Game*, published by iUniverse, Inc. The book won the 2010 Editor's Choice Award from iUniverse for its creative representation of the linkage between mathematics and sports excellence. Interested readers are encouraged to refer to that book for extensive computational examples, the details of which are not provided here. Soccer math for kicks is a good pun to cite here to inspire sports-minded readers.

MATH-MY-PHYSICS

There is no physics without math. Math-my-physics is my idea of collating cool mathematical relationships between math and physics. Physics is a natural science that deals with the structure of matter. Everything we know is a matter. Physics explains the interactions between matters and their fundamental elements. Physics explains how things move in space with respect to time. Physics operates within the realms of energy, force, motion, light, electricity, radiation, gravity, and other physical phenomenon. All the theories and applications of physics are predicated on mathematical relationships. All the laws of physics are expressed in mathematical equations. Math-my-physics makes this relationship clear in a fun way.

MATH-MY-MORNING

There are many ways to have fun with math. In July 28, 2022 posting by New York Post, it was revealed that a British mathematician has come up with the secret to the perfect morning routing (https://nypost.com/2022/07/28/secret-to-the-perfect-morning-routine-revealed-by-mathematician/). The optimal morning routine is allegedly, mathematically, proven to improve one's day. The UK mathematician claims to have devised a scientific formula for the best way to begin the day. The claim says: "Having this formula is a great tool to help start the day right," British Science Association head Dr. Anne-Marie Imafidon told The Sun newspaper of her carpe diem calculator. She invented this sunrise-reaping recipe after analyzing a survey of 2,000 UK residents on the optimal morning routine. The poll, commissioned by Special K Crunchy Oat Granola, found that 6:44 a.m. is the ideal wake-up time, while it's best to spend 10 minutes in the shower. Meanwhile, people should allot 18 minutes to eating breakfast and 21 minutes to exercise, per the study. To calculate if one has the optimal morning routine, the scientist prescribes doubling the amount of time allocated to eating breakfast, then adding that to how many minutes are spent on showers and exercise. The fun never ends with mathematicians. What will they think of next?

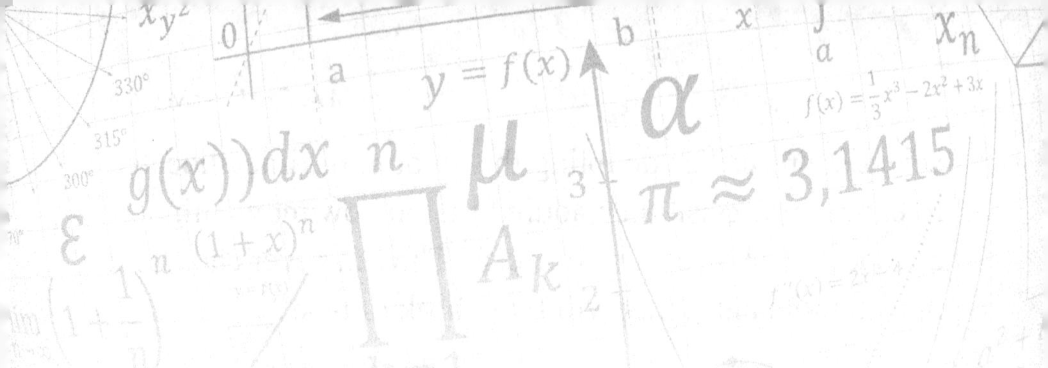

MATH MAGIC OF MATH MAGICIANS

There are countless online resources to inspire Mathmamiya reactions. In this section, some common as well as uncommon ones are listed. Readers are encourage to seek them out based on interest and time.

Cubes and higher powers.

Square Numbers.

Prime Numbers.

Divisibility rules.

Fractions, decimals, and percents.

Order of operations.

Roman Numerals.

Basic conversions.

FOIL method.

Square a 2-digit number.

Double and half method.

Square a number ending in 5.

Square a number from 40-49.

Square a number from 50-59.

Square a number from 90-99.

Multiply two numbers whose digits add to 10.

Multiply numbers less than 100.

Multiply numbers greater than 100.

Multiply numbers in the form of ab + bc.

Multiply by 5.

Multiply by 11.

Multiply by 25.

Multiply by 50.

Multiply by 101.

Add and subtract fractions. Compare fractions.

Multiply mixed numbers.

Add sequences in the form of 1+2+ . . .+n.

Add sequences in the form of 1+3+ . . . 2n-1.

Percents.

GCD.

LCM.

Definitions.

Ratios.

Square roots.

Subtracting reverses.

Approximating series.

Square a number ending in 6.

Square a number ending in 7.

Square a number ending in 8.

Square a number ending in 9.

Differences of 2 squares.

Multiply two numbers ending in 5.

Multiply numbers whose 10's digits add to 10.

Add squared numbers in the form $a^2 + (2a)^2$

Add squared numbers in the form of $a^2 + (3a)^2$

Add squared numbers in the form of $a^2 + (7a)^2$

Add squared numbers in the form of $a^2 + (10a)^2$

Multiply numbers less than 1,000.

Multiply by 15.

Multiply by 75.

Multiply by 111.

Multiply by 1001.

Multiply by 1111.

Multiply by 9.

Multiply by 8.

Divide by 40.

Divide by 25.

Divide by 9.

Multiply two numbers that Add to 5.

Multiply two numbers with the same Ten's.

Multiply one number above and below 100.

Add two Squared Numbers.

Multiply by 286.

Multiply by 429.

Multiply by 572.

Multiply by 715.

Multiply by 858.

Multiply by 3367.

Multiply by 14443.

Multiply by 101101.

Multiply by 142857.

Changing Probability To Odds.

Permutations.

Combinations.

Probability with Coins.

Probability with Dice.

Probability with Cards.

Probability with Balls.

Pythagorean Triples.

Inscribed Circles.

Circumscribed Circles.

Imaginary Numbers.

Add three Fractions.

MATH CONVERSION FACTORS

Mathematical conversions from one space, time, and dimension into another are essential for many of the technological advancements that we enjoy today. This section presents selected useful conversion factors. Conversions of distance, weight, volume, and so on affect everything we do at home, at work, in commerce, and at leisure.

Kilometer-Mile Conversions			
Kilometers to	**Miles**	**Miles to**	**Kilometers**
1	0.6	1	1.6
5	3.1	5	8.05
10	6.2	10	16.0
20	12.4	20	32.1
30	18.6	30	48.2
40	24.8	40	64.3
50	31.1	50	80.5
60	37.3	60	96.6

70	43.5	70	112.7
80	49.7	80	128.7
90	55.9	90	144.8
100	62.1	100	160.9
500	310.7	500	804.7
1,000	621.4	1,000	1609.3

Metric Tables

Capacity		**Area**	
10 milliliters	= 1 centiliter	100 sq. millimeters	= 1 sq. centimeter
10 centiliters	= 1 deciliter	100 sq. centimeters	= 1 sq. decimeter
10 deciliters	= 1 liter	100 sq. decimeters	= 1 sq. meter (centare)
10 liters	= 1 dekaliter	100 sq. meters	= 1 are
10 dekaliters	= 1 hectoliter	10,000 sq. meters	= 1 hectare
1,000 liters	= 1 kiloliter (stere)	100 hectares	= 1 sq. kilometer
LENGTH		WEIGHT	
10 millimeters	= centimeter (cm)	10 milligrams	= 1 centigram
10 centimeters	= 1 decimeter	10 centigrams	= 1 decigram
10 decimeters	= 1 meter (m)	10 decigrams	= 1 gram
10 meters	= 1 dekameter	1,000 grams	= 1 kilogram (kilo)
100 meters	= 1 hectometer	100 kilograms	= 1 quintal
1,000 meters	= 1 kilometer	1,000 kilograms	= 1 metric ton

Metric Equivalent Of U.S. Weights And Measures

DRY MEASURE		LONG MEASURE	
1 Pint	= .550599 liter	1 inch	= 2.54 centimeters
1 quart	= 1.101197 liters	1 yard	= .914401 meter
1 peck	= 8.80958 liters	1 mile	= 1.609347 kilometers
1 bushed	= .35238 hectoliter		

LIQUID MEASURE		SQUARE MEASURE	
1 Pint	= .473167 liter	1 sq. inch	6.4516 sq. centimeters
1 quart	= .946332 liter	1 sq. foot	9.29034 sq. decimeters
1 gallon	= 3.785329 liters	1 sq. yard	.836131 sq. meter
		1 acre	.40469 hectares
		1 sq. mile	2.59 sq. kilometers
		1 sq. mile	259 hectares
AVOIRDUPOIS MEASURE		**CUBIC MEASURE**	
1 ounce	= 28.349527 grams	1 cu. Inch	= 16.3872 cu. Centimeters
1 pound	= .453592 kilograms	1 cu. Foot	= .028317 cu. Meter
1 short ton	= .90718486 metric ton	1 cu. yard	= .76456 cu meter
1 long ton	= 1.01604704 metric tons		

Useful Mathematical Relationships

$$\sin\theta = \frac{b}{c} \qquad \csc\theta = \frac{c}{b}$$

$$\sin\theta = \frac{a}{c} \qquad \sec\theta = \frac{c}{a}$$

$$\tan\theta = \frac{b}{a} \qquad \cot\theta = \frac{a}{b}$$

1 radian	= 57.3°
1 inch	= 2.54 cm
1 gallon	= 231 in^3
1 kilogram	= 2.205 lb
1 newton	= 1 kg • m/s^2
1 joule	= 1 N • m
1 watt	= 1 J/s
1 pascal	= 1 N/m^2

1 BTU	= 778 ft-lb
	= 252 cal
	= 1,054.8 J
1 horsepower	= 745.7 W
1 atmosphere	= 14.7 lb/in^2
	= 1.01 • 10^5 N/m^2

Mathematical Constants

Speed of light:	2.997,925 × 10^{10} cm/sec
	983.6 × 10^6 ft/sec
	186,284 miles/sec
Velocity of sound:	340.3 meters/sec
	1116 ft/sec
Gravity:	9.80665 m/sec square
Acceleration:	32.174 ft/sec square
	386.089 inches/sec square

Area Relationships

Multiply	by	to obtain
acres	43,560	sq feet
	4,047	sq meters
	4,840	sq yards
	0.405	hectare
sq cm	0.155	sq inches
sq feet	144	sq inches
	0.09290	sq meters
	0.1111	sq yards
sq inches	645.16	sq millimeters
sq kilometers	0.3861	sq miles

sq meters	10.764	sq feet
	1.196	sq yards
sq miles	640	acres
	2.590	sq kilometers

Volume Relationships

Multiply	by	to obtain
acre-foot	1233.5	cubic meters
cubic cm	0.06102	cubic inches
cubic feet	1728	cubic inches
	7.480	gallons (US)
	0.02832	cubic meters
	0.03704	cubic yards
liter	1.057	liquid quarts
	0.908	dry quarts
	61.024	cubic inches
gallons (US)	231	cubic inches
	3.7854	liters
	4	quarts
	0.833	British gallons
	128	U.S. fluid ounces
quarts (US)	0.9463	liters

Rule of Arithmetic Operations

BODMAS (Brackets, Off, Division, Multiplication, Addition, Subtraction)

PEMDAS (Parenthesis, Exponent Multiply or Divide which ever comes first)

Energy And Heat Power Relationships

Multiply	by	to obtain
BTU	1055.9	joules
	0.2520	kg-calories
watt-hour	3600	joules
	3.409	BTU
HP (electric)	746	watts
BTU/second	1055.9	watts
watt-second	1.00	joules

Mass Relationships

Multiply	by	to obtain
carat	0.200	cubic grams
grams	0.03527	ounces
kilograms	2.2046	pounds
ounces	28.350	grams
pound	16	ounces
	453.6	grams
stone (UK)	6.35	kilograms
	14	pounds
ton (net)	907.2	kilograms
	2000	pounds
	0.893	gross ton
	0.907	metric ton
ton (gross)	2240	pounds
	1.12	net tons
	1.016	metric tons
tonne (metric)	2,204.623	pounds
	0.984	gross pound
	1000	kilograms

Temperature Relationships

Conversion formulas	
Celsius to Kelvin	K = C + 273.15
Celsius to Fahrenheit	F = (9/5)C + 32
Fahrenheit to Celsius	C = (5/9)(F – 32)
Fahrenheit to Kelvin	K = (5/9)(F + 459.67)
Fahrenheit to Rankin	R = F + 459.67
Rankin to Kelvin	K = (5/9)R

Velocity Relationships

Multiply	by	to obtain
feet/minute	5.080	mm/second
feet/second	0.3048	meters/second
inches/second	0.0254	meters/second
km/hour	0.6214	miles/hour
meters/second	3.2808	feet/second
	2.237	miles/hour
miles/hour	88.0	feet/minute
	0.44704	meters/second
	1.6093	km/hour
	0.8684	knots
knot	1.151	miles/hour

Pressure Relationships

Multiply	by	to obtain
atmospheres	1.01325	bars
	33.90	feet of water
	29.92	inches of mercury
	760.0	mm of mercury

bar	75.01	cm of mercury
	14.50	pounds/sq inch
dyne/sq cm	0.1	N/sq meter
newtons/sq cm	1.450	pounds/sq inch
pounds/sq inch	0.06805	atmospheres
	2.036	inches of mercury
	27.708	inches of water
	68.948	millibars
	51.72	mm of mercury

Distance Relationships

Multiply	by	to obtain
angstrom	10^{-10}	meters
feet	0.30480	meters
	12	inches
inches	25.40	millimeters
	0.02540	meters
	0.08333	feet
kilometers	3280.8	feet
	0.6214	miles
	1094	yards
meters	39.370	inches
	3.2808	feet
	1.094	yards
miles	5280	feet
	1.6093	kilometers
	0.8694	nautical miles
millimeters	0.03937	inches
nautical miles	6076	feet
	1.852	kilometers
yards	0.9144	meters
	3	feet
	36	inches

Common Mathematical Notations

UNITS OF MEAS.	ABBREV.	RELATION	UNITS OF MEAS.	ABBREV.	RELATION
meter	m	length	degree Celsius	°C	temperature
hectare	ha	area	Kelvin	K	thermodynamic temp.
tonne	t	mass	pascal	Pa	pressure, stress
kilogram	kg	mass	joule	J	energy, work
nautical mile	M	distance (navigation)	Newton	N	force
knot	kn	speed (navigation)	watt	W	power, radiant flux
liter	L	volume or capacity	ampere	A	electric current
second	s	time	volt	V	electric potential
hertz	Hz	frequency	ohm	Ω	electric resistance
candela	cd	luminous intensity	coulomb	C	electric charge

Kitchen Measurements

A pinch	1/8 tsp or less
3 tsp	1 tbsp
2 tbsp	1/8 c
4 tbsp	1/4 c
16 tbsp	1 c
5 tbsp + 1 tsp	1/3 c
4 oz	1/2 c
8oz	1 c
16 oz	1 lbs
1 oz	2 tbsp fat or liquid
1 c of liquid	1/2 pt
2 c	1 pt
2 pt	1 qt
4 c of liquid	1 qt
4 qts	1 gallon
8 qts	1 peck (such as apples, pears, etc)
1 jigger	1 ½ floz
1 jigger	3 tbsp.

MATH CLOSURE

We have now reached the end of my mathmamiya presentation. Like in closing arguments in a court of law, I present to you that math underpins everything. Essentially, if there is a math proof, the argument has been proven.

To bring this edition of the book to closure, I, hereby, reiterate that this is not a book to teach mathematics. Rather, it is a book to tease mathematics to whet the appetite of readers for more. Please stay tuned for future Mathmamiya fun books from ABICS Publications. Visit www.ABICSPublications.com for a listing of the diverse titles available in the ABICS books series.

Appendix

Sites to See

Websites for math fun and facts

http://www.kidsmathgamesonline.com/pictures/shapes/hexagon.html

www.PhysicsofSoccer.com

https://www.mathisfun.com

https://www.ixl.com/

https://www.adventureacademy.com/

https://www.education.com/resources/math

https://cathyduffyreviews.com/

https://www.weareteachers.com/

https://www.temu.com/

https://www.testgorilla.com/

https://brilliant.org/

https://www.astrologiez.com/

https://www.coolmath4kids.com/

https://www.groupon.com/deals/mathnasium